NATURE'S REVENGE

TWISTING TORNADOES

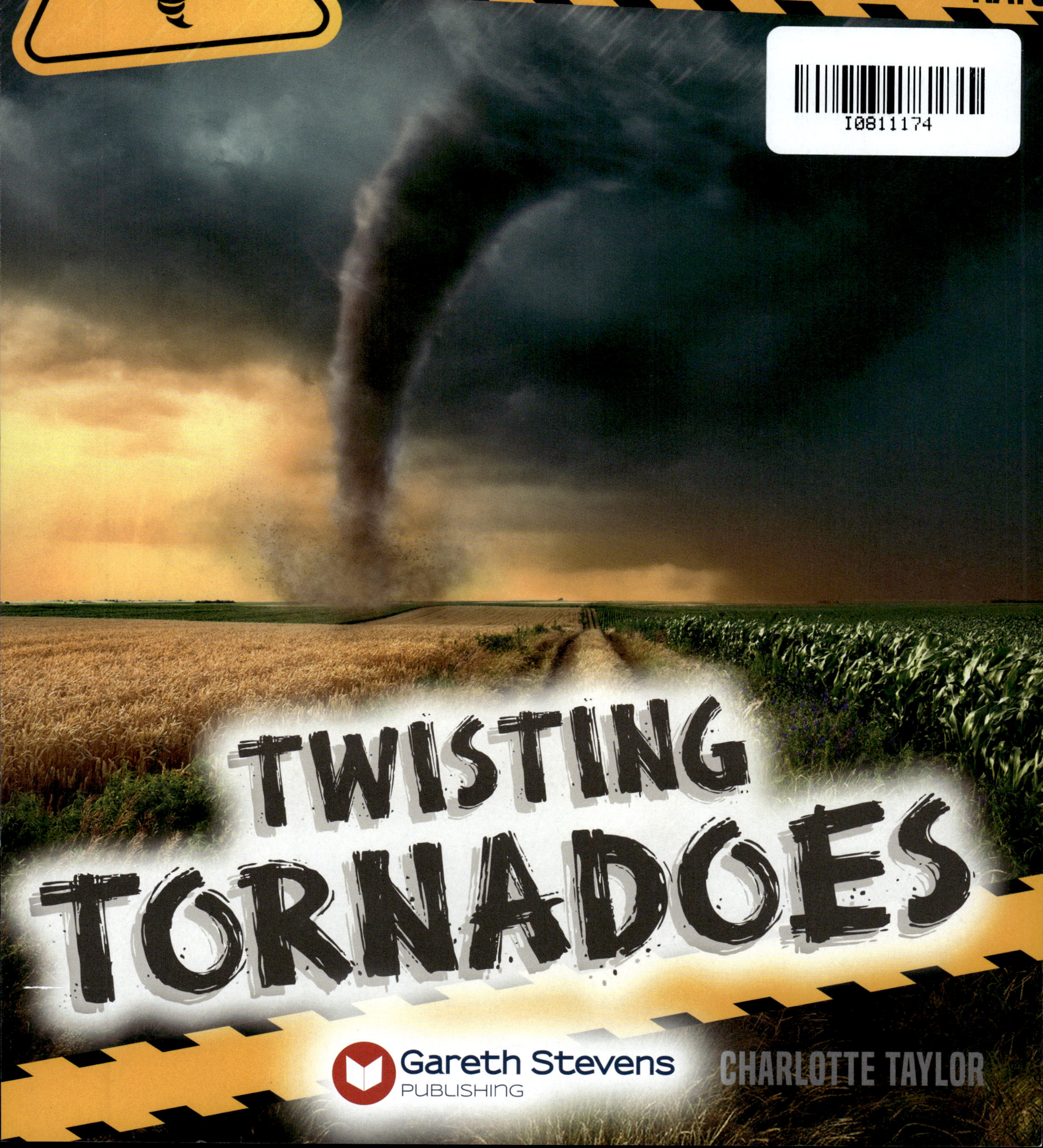

Gareth Stevens PUBLISHING

CHARLOTTE TAYLOR

Please visit our website, www.garethstevens.com. For a free color catalog of all our high-quality books, call toll free 1-800-542-2595 or fax 1-877-542-2596.

Cataloging-in-Publication Data
Names: Taylor, Charlotte, 1978-.
Title: Twisting tornadoes / Charlotte Taylor.
Description: New York : Gareth Stevens Publishing, 2023. | Series: Nature's revenge | Includes glossary and index.
Identifiers: ISBN 9781538280591 (pbk.) | ISBN 9781538280614 (library bound) | ISBN 9781538280607 (6pack) | ISBN 9781538280621 (ebook)
Subjects: LCSH: Tornadoes–Juvenile literature. | Natural disasters–Juvenile literature.
Classification: LCC QC955.2 T39 2023 | DDC 551.55'3–dc23

Portions of this work were originally authored by Elizabeth Krajnik and published as *Twisted by Tornadoes.* All new material in this edition was authored by Charlotte Taylor.

Published in 2023 by
Gareth Stevens Publishing
29 East 21st Street
New York, NY 10010

Designer: Leslie Taylor
Editor: Megan Quick

Photo credits: Cover, p. 1 (tornado photo) Rasica/Shutterstock.com; pp. 1–32 (icons-series artwork) Vector by/Shutterstock.com; pp. 1–32 (hazard tape-series artwork) DDevecee/Shutterstock.com; p. 5 agefotostock/Alamy.com; p. 7 Designua/Shutterstock.com; p. 9 (top) https://commons.wikimedia.org/wiki/File:FEMA_-_3747_-_Photograph_by_Andrea_Booher_taken_on_05-04-1999_in_Oklahoma.jpg; p. 9 (bottom) https://commons.wikimedia.org/wiki/File:FEMA_-_3751_-_Photograph_by_Andrea_Booher_taken_on_05-04-1999_in_Oklahoma.jpg; p. 11 Aaron of L.A. Photography/Shutterstock.com; p. 12 (chart) https://www.weather.gov/hun/efscale_explanation; p. 13 Newtonian/Shutterstock.com; p. 15 (top) Ryan McGinnis/Alamy.com; p. 15 (bottom) imac/Alamy.com; p. 17 GSW Photography/Shutterstock.com; p. 19 Photoguru73/Shutterstock.com; p. 21 Associated Press/APImages.com; p. 23 Gino Santa Maria/Shutterstock.com; p. 25 Wynn photography/Shutterstock.com; p. 27 trgrowth/Shutterstock.com; p. 29 Todd Shoemake/Shutterstock.com.

Printed in the United States of America

CPSIA compliance information: Batch #CSGS23: For further information contact Gareth Stevens, New York, New York at 1-800-542-2595.

CONTENTS

Words in the glossary appear in **bold** type the first time they are used in the text.

FIERCE WINDS

A tornado is one of the most powerful and **dangerous** weather events on Earth. Its winds can destroy homes, lift cars into the air, and knock down trees and power lines. Some of the strongest tornadoes have wiped out entire towns. People who are in the path of a tornado need to act fast to stay safe.

The United States has more tornadoes than any other country—about 1,200 a year. Certain areas are more likely to see tornadoes than others, but they have happened in all 50 states, so everyone should know how to prepare. Knowing where and when a tornado will strike is tricky, but better **technology** is helping scientists **predict** these deadly events and save lives.

A tornado moves through a field in Kansas.

Nature's Fury

A tornado is a type of natural disaster. This is an **extreme** event that happens in nature and causes a great deal of **damage**. It can destroy buildings, knock out power, and hurt or kill people. Besides tornadoes, natural disasters include **hurricanes**, floods, and **earthquakes**.

CREATING A TORNADO

You've seen pictures of funnel clouds, but what exactly is a tornado? Many start off as thunderstorms in cumulonimbus clouds, which are very tall clouds. Two types of air come together: warm, moist air from the Gulf of Mexico mixes with cool, dry air from Canada. This makes the atmosphere unstable. Changes in the direction and speed of the wind cause these air masses to swirl.

Next, the air near the ground pushes up against the air from the cloud to create a funnel. The funnel is a spinning column of air that sucks up more warm air from the ground. Then the funnel becomes longer and stretches from the cloud to the ground. When the funnel touches the ground, it's considered a tornado.

Supercells

Not all thunderstorms can create tornadoes. The ones that are most likely to cause a tornado are called supercell thunderstorms. They have the tall clouds and strong upward moving winds (called updrafts) that create tornadoes. Supercells often include hail and heavy winds. They are large, long-lasting storms.

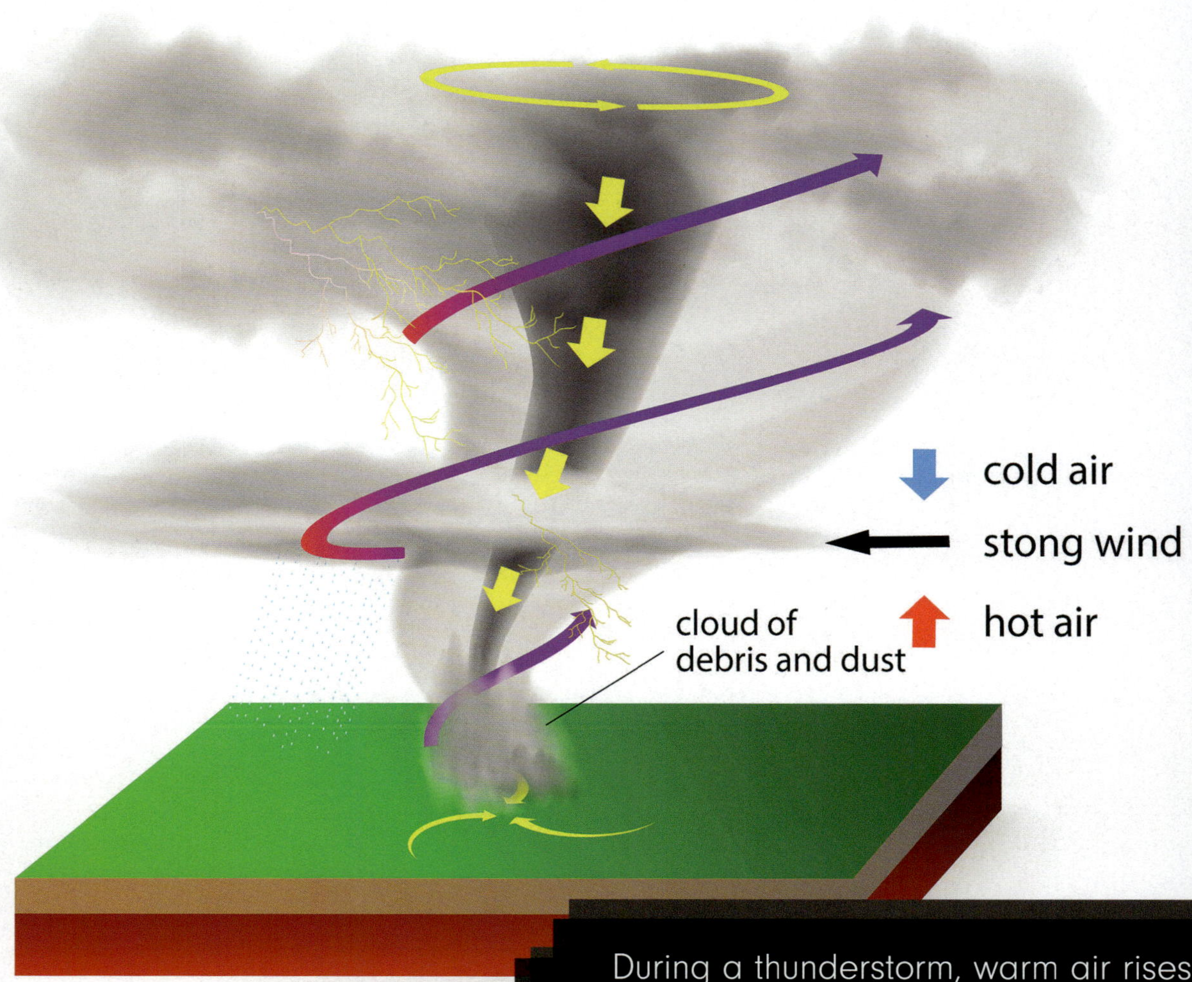

During a thunderstorm, warm air rises up as cold air pushes down. The high winds begin to spin, forming a funnel.

WHAT TO EXPECT

Tornadoes usually happen between midafternoon and evening, and they most often occur between March and August. But they can happen at any time of day, any time of year. Warning signs for a tornado include dark, greenish skies and a large cloud that's low in the sky. There may also be hail and a loud sound like a train coming toward you.

A tornado can cause damage to an area that's more than 1 mile (1.6 km) wide and 50 miles (80.5 km) long. However, the funnel itself may be much smaller. After touching down, tornadoes usually move from west to east at about 28 miles (45.1 km) per hour. While most only touch down for about five minutes, some can last an hour or more.

Residents walk through the remains of their homes after a tornado hit Midwest City, Oklahoma, in 1999.

Deadly Gusts

The highest tornado wind speeds ever measured were part of a cluster of tornadoes that touched down in Oklahoma and Kansas on May 3, 1999. An amazing 74 tornadoes hit the area in less than one day. One of those tornadoes had winds that reached just over 300 miles (482.8 km) per hour. In all, 46 people were killed and 800 were hurt.

HOW STRONG IS IT?

Meteorologists are scientists who study weather, climate, and the atmosphere. They use a rating system to describe how strong a tornado is. The Fujita scale (F-scale) was developed by Dr. Tetsuya Theodore Fujita in 1971. The scale was based on the damage a tornado caused, from light damage such as broken tree branches to "**incredible**" damage like houses being moved and destroyed.

Then, in the late 1990s, meteorologists started looking for ways to improve the F-scale. They studied the damage caused by the 1999 Oklahoma tornadoes. They found the F-scale was not accurate when it **estimated** tornado wind speeds at higher levels. The F-scale was updated to become the Enhanced Fujita scale (EF-scale).

Trees were ripped out of the ground by a tornado in Boscobel, Wisconsin, in 2021. Scientists study this kind of damage to figure out how to rate a tornado.

How the Scales Work

Both the F-scale and the EF-scale depend on speed estimates for 3-second wind gusts based on damage caused. These estimates aren't actual wind-speed measurements. Tornado winds are so strong that they would destroy the tools needed to measure their wind speeds.

The new EF-scale gives detailed descriptions of damage to 23 types of buildings and five other types of objects. Different buildings include barns, shopping malls, and motels. The amount of damage to each kind of building is examined and used in figuring wind speeds.

A tornado's rating is given by the National Weather Service (NWS). The amount of damage is figured and the highest wind speed is estimated. The rating is then assigned, anywhere from EF-0 to EF-5. An EF-0 rating means gusts reached between 65 and 85 miles (105 and 137 km) per hour. An EF-5 rating includes tornadoes with gusts over 200 miles (322 km) per hour.

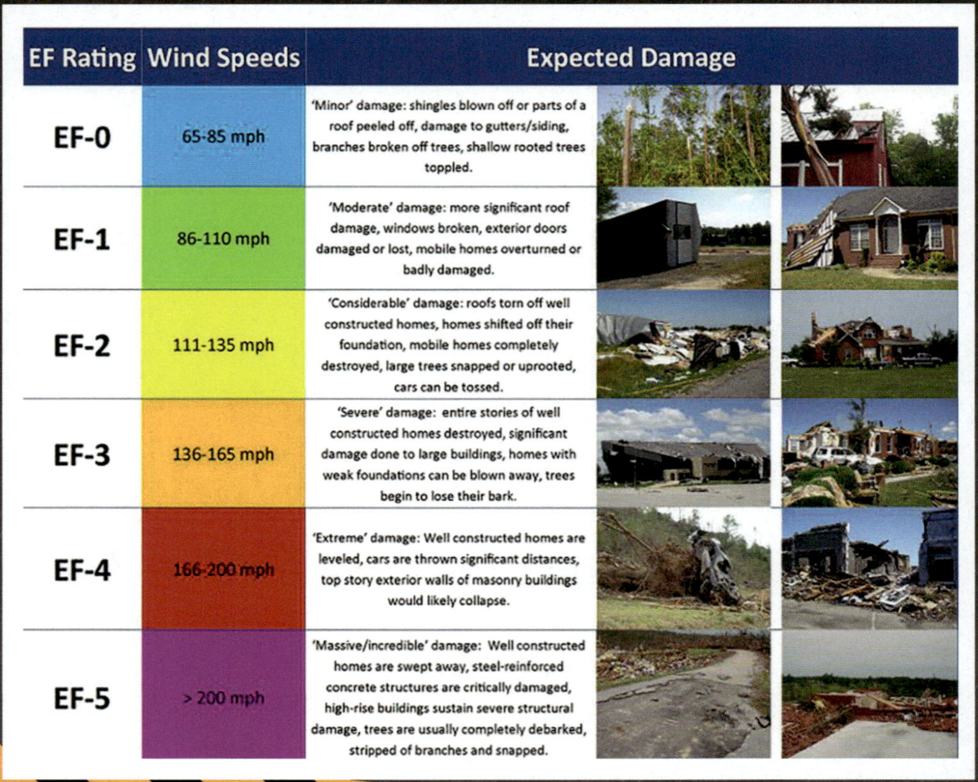

EF Rating	Wind Speeds	Expected Damage
EF-0	65-85 mph	'Minor' damage: shingles blown off or parts of a roof peeled off, damage to gutters/siding, branches broken off trees, shallow rooted trees toppled.
EF-1	86-110 mph	'Moderate' damage: more significant roof damage, windows broken, exterior doors damaged or lost, mobile homes overturned or badly damaged.
EF-2	111-135 mph	'Considerable' damage: roofs torn off well constructed homes, homes shifted off their foundation, mobile homes completely destroyed, large trees snapped or uprooted, cars can be tossed.
EF-3	136-165 mph	'Severe' damage: entire stories of well constructed homes destroyed, significant damage done to large buildings, homes with weak foundations can be blown away, trees begin to lose their bark.
EF-4	166-200 mph	'Extreme' damage: Well constructed homes are leveled, cars are thrown significant distances, top story exterior walls of masonry buildings would likely collapse.
EF-5	> 200 mph	'Massive/incredible' damage: Well constructed homes are swept away, steel-reinforced concrete structures are critically damaged, high-rise buildings sustain severe structural damage, trees are usually completely debarked, stripped of branches and snapped.

Rows of destroyed homes lined the streets of Moore, Oklahoma, after an EF-5 tornado hit in 2013.

EF-5 Hot Spot

EF-5 tornadoes are very rare, but Moore, Oklahoma, has experienced two of them, in 1999 and 2013. These tornadoes caused severe damage to the area, with 60 deaths in total. The city has also seen at least 16 other tornadoes since the mid-1970s. Scientists are not sure why Moore has been hit so much—they say it may just be bad luck.

TORNADO FORECAST

Some tornadoes form quickly and can't be predicted. But meteorologists are skilled at figuring out when conditions are right for a tornado to form. They use different kinds of technology to locate and track thunderstorms that might turn into tornadoes.

One helpful type of technology is Doppler **radar**. Doppler radar determines the rate of **precipitation** and the speed and direction of the wind. Computer programs read and explain data from the Doppler radar so meteorologists can keep an eye out for possible tornadoes. Scientists have also started working with phased array technology, which can scan an entire storm in a minute to see if tornadoes are likely.

A **mobile** Doppler radar (above) is used in Iowa to detect possible tornadoes. It can look at a storm to gather information that will help scientists better understand tornadoes.

A More Dangerous Time

A century ago, people only knew a tornado was coming when they saw it or heard it. This meant that they had almost no time to prepare, and many people died. In 1925, the Tri-State Tornado traveled through parts of Missouri, Illinois, and Indiana. It killed 695 people, making it the deadliest tornado in U.S. history.

WATCHES AND WARNINGS

Once a dangerous storm is spotted, people need to be warned as soon as possible. Giving people notice that a tornado could develop can save lives. The NWS issues a tornado watch when a tornado is possible. A tornado warning is issued when someone has seen a tornado touch down in the area or tools that detect, or find, tornadoes have suggested a tornado is very likely to develop.

Many towns and cities where tornadoes are common also use warning sirens to alert people that a tornado has been spotted. Once a tornado warning is issued, the loud siren goes off to let people know that they should move to shelter immediately.

A tornado warning will be sent out if meteorologists spot a supercell storm like this one, since there is a high chance that a tornado will form.

Short Notice

Unlike snowstorms or hurricanes, tornado warnings give people minutes rather than days to prepare. In fact, the NWS estimates the time between a warning and the moment a tornado touches down averages between 8 and 11 minutes. This may not seem like much time, but often it's enough to get to safety.

GET READY

If you want to be ready for a tornado, do not wait until a warning is issued. Plan ahead! A great first step is to make an emergency kit. This could include bottled water, food that won't spoil, medicine, a first aid kit, a whistle to alert rescuers, a battery-powered radio, a flashlight, and extra batteries.

Once you have your kit together, talk with your family about your tornado plan. Decide where you will shelter, like a basement, storm cellar, or windowless first-floor room. Then practice your plan. Know how to get weather updates from a local news station or NOAA Weather Radio. You can also sign up for local notifications on a cell phone.

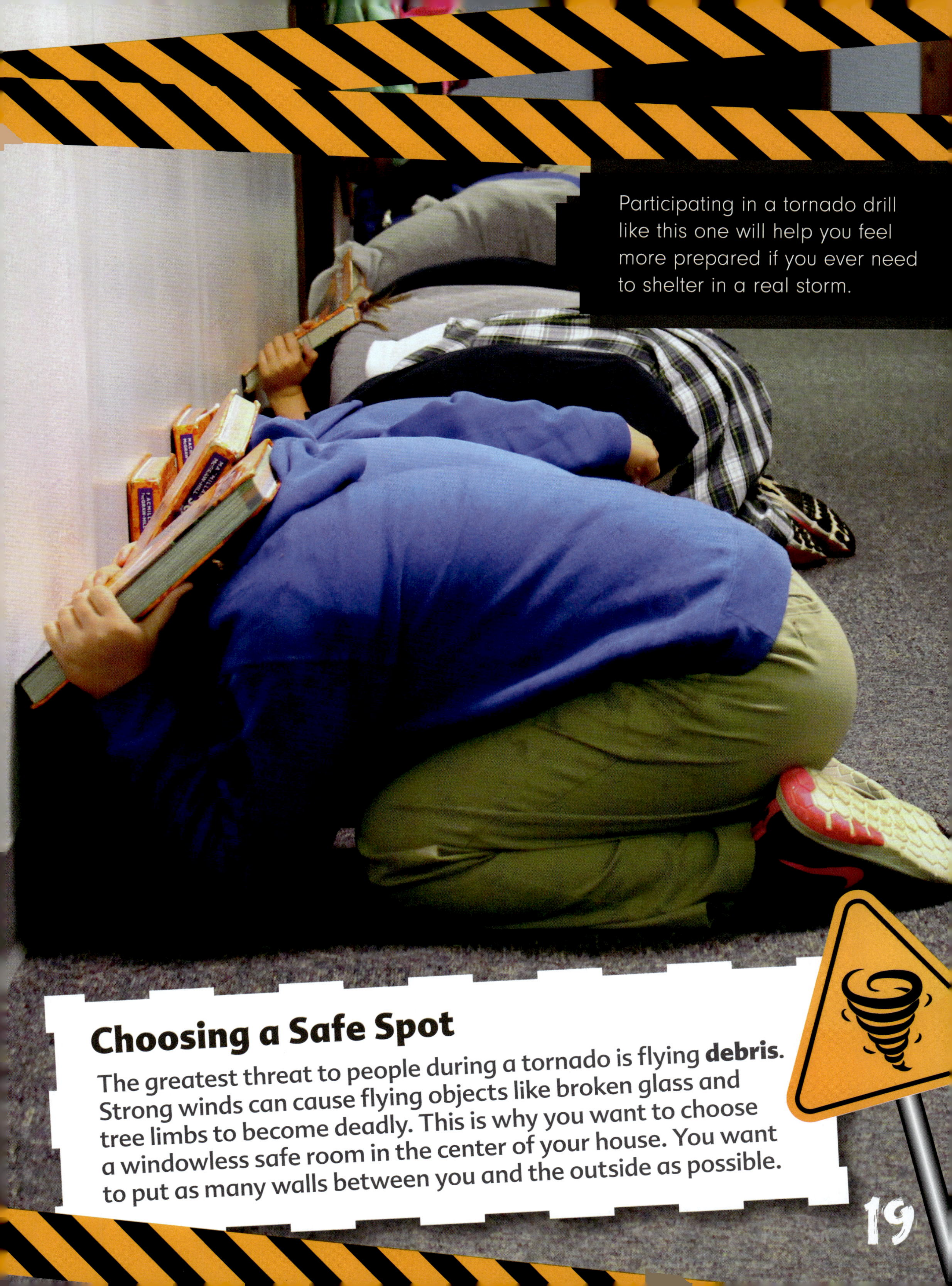

Participating in a tornado drill like this one will help you feel more prepared if you ever need to shelter in a real storm.

Choosing a Safe Spot

The greatest threat to people during a tornado is flying **debris**. Strong winds can cause flying objects like broken glass and tree limbs to become deadly. This is why you want to choose a windowless safe room in the center of your house. You want to put as many walls between you and the outside as possible.

TAKE SHELTER

Imagine that you have just heard a tornado warning on the radio. Take shelter immediately. If you are home, go to your safe room. If you are someplace like school or a shopping mall, don't panic. Most places have a plan in place for a natural disaster. Head to the building's tornado shelter, basement, or a windowless inner room. Sit on the floor and cover your head with your arms.

If you're outside, finding safety is more difficult. A car may be the safest place. Try to have an adult drive you to the nearest shelter. If there's no shelter nearby, stay in the car, put on your seatbelt, and cover your head with your arms and anything else you can find to keep your head safe.

Mobile Homes and Tornadoes

One place that is not safe in a tornado is a mobile home. A mobile home is not built to handle high winds and it is usually not anchored well to the ground. If there is no other shelter nearby, it is safer to leave the mobile home and lie in a low, open area while covering your head.

A 2020 tornado in Mississippi destroyed this entire home except for a small safe room where the family took shelter.

SUPPORT FOR SURVIVORS

Once a tornado has passed, groups of people spring into action to help. Neighbors check on neighbors, and emergency workers look for people trapped in fallen buildings. The Red Cross is one of many groups that provide shelter and food to survivors of natural disasters. It also helps with cleanup by handing out items such as shovels, rakes, and work gloves. The Red Cross helps families get in touch with each other after a natural disaster.

Many smaller groups also come to the aid of tornado survivors. They help by giving money to larger relief organizations, collecting and handing out supplies, helping with cleanup, and helping in hospitals and clinics. It takes many people to help communities recover from a natural disaster.

Volunteers help clean up after tornadoes ripped through the Saint Louis, Missouri, area in 2011.

Trapped in the Rubble

If you are in a tornado and become trapped, move as little as possible to avoid kicking up dust. Use the whistle from your emergency kit or tap on a pipe or wall to alert a rescue team. Once you make it outside, watch out for debris and fallen power lines.

RUIN AND RECOVERY IN JOPLIN

One of the most destructive tornadoes in U.S. history happened on May 22, 2011. The EF-5 tornado ripped through Joplin, Missouri, killing 161 people and injuring many others. The tornado was 0.5 mile (0.8 km) wide and grew to 0.75 mile (1.2 km) wide. It traveled about 22 miles (20.9 km) with winds of about 200 miles (321.9 km) per hour. Homes, buildings, and cars were destroyed. The cost of damages was almost $3 billion.

Since the tornado struck, Joplin has added about five new houses a week (more than 2,500 new houses as of 2021). Its hospital, parks, public library, and schools have been rebuilt. Many of the new buildings have tornado shelters.

The Joplin tornado damaged or destroyed more than 7,400 homes.

Joplin Hope

After the Joplin disaster, residents came together to help each other rebuild. They now hope that people in other areas can learn from their experiences. The Joplin Hope Center for Disaster Recovery was created to share information about the tornado and to help communities prepare for and cope with similar disasters.

MORE TORNADOES?

Our planet's climate is changing. It is getting hotter, causing glaciers to melt and sea levels to rise. This is called global warming. Rising sea levels lead to more moisture in the air, which can cause more severe weather. Does that mean we will see more tornadoes in the future? Not necessarily, scientists say.

Warm, moist air is one element in forming a tornado. But another important part, shifting winds, may actually decrease with global warming. So it's possible we'll have fewer tornadoes as the planet heats up. So far, there has not been any major increase or decrease in the number of tornadoes each year. Scientists need to learn more before they know if that will change.

Global Warming Basics

There are many reasons why the planet is getting warmer. Some we cannot control, like changes in the sun. But people play an important part as well. When we drive, cook, or heat our homes, we use **fossil fuels** like coal and oil. These fuels send gases into the air. The gases trap the sun's heat. Over time, Earth gets warmer.

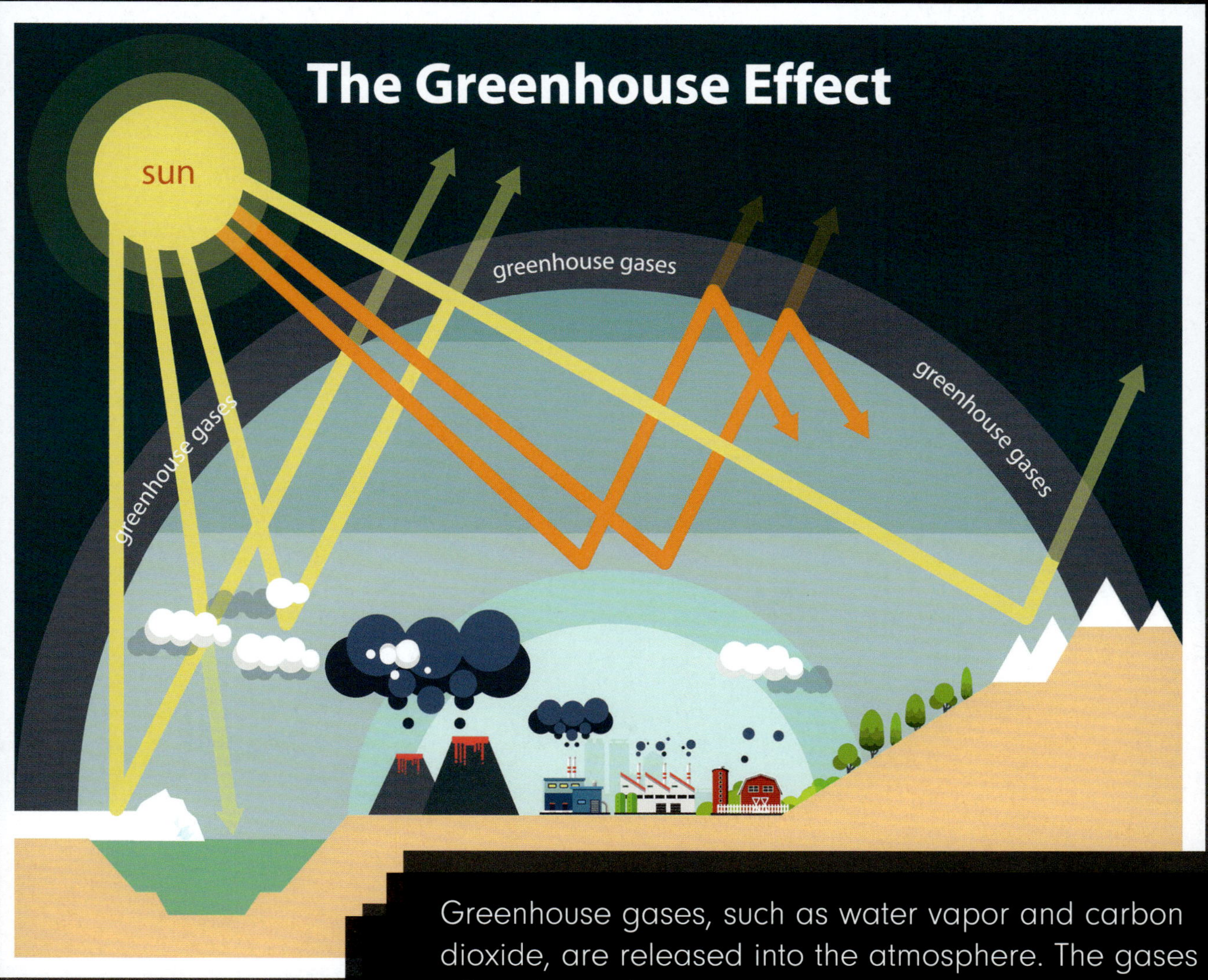

Greenhouse gases, such as water vapor and carbon dioxide, are released into the atmosphere. The gases allow the sun's light to leave the atmosphere, but they trap the heat. This is called the greenhouse effect.

TOMORROW'S STORMS

Tornadoes are some of the most dangerous events in nature. Scientists are working on making more exact tornado predictions, but deadly storms can still surprise us. No matter where you live, it is smart to be prepared and plan ahead in order to stay safe.

We may not know whether global warming will cause more tornadoes in the long run. We do know that there is a good chance that extreme weather will get worse in years to come. People have played a role in causing climate change, and they can also play a role in slowing it down. It's not too late to make our world safer and healthier.

Hi-Tech Tornado Tools

Meteorologists are developing new technology to warn people earlier of a possible tornado. The Warn-on-Forecast system uses computer models to study weather data and predict extreme events before they happen. This would allow meteorologists to issue warnings hours, rather than minutes, before the event. When it is fully developed, this technology could save many lives.

A tornado touches down in Colorado. In the future, meteorologists may be able to know when a tornado is coming before it even forms.

GLOSSARY

damage: harm

dangerous: unsafe

debris: the remains of something that has been broken

earthquake: a shaking of the ground caused by the movement of Earth's crust

estimate: to make a guess based on known facts

extreme: great or severe

fossil fuel: matter formed over millions of years from plant and animal remains that is burned for power

hurricane: a powerful storm that forms over water and causes heavy rainfall and high winds

incredible: hard to believe

mobile: moving

precipitation: water that falls to the ground as hail, mist, rain, sleet, or snow

predict: to say what will or might happen in the future

radar: a machine that uses radio waves to locate and identify objects

technology: the use of science to invent useful things or to solve problems

FOR MORE INFORMATION

BOOKS

Drimmer, Stephanie. *National Geographic Kids Ultimate Weatherpedia.* Washington, DC: National Geographic Kids, 2019.

Gibbons, Gail. *Tornadoes!* New York, NY: Holiday House, 2019.

Tarshis, Lauren. *I Survived True Stories: Tornado Terror.* New York, NY: Scholastic, 2017.

WEBSITES

Ducksters: Tornadoes
www.ducksters.com/science/earth_science/tornadoes.php
Check out more fun facts about tornadoes.

National Geographic Kids: What Is Climate Change?
www.natgeokids.com/uk/discover/geography/general-geography/what-is-climate-change/
Find out all about the causes and effects of climate change and how you can help.

Weather Wiz Kids: Tornadoes
www.weatherwizkids.com/weather-tornado.htm
Learn more about the different parts of a tornado and try out a tornado experiment.

INDEX